PLANTS with SUPERPOWERS

PLANTS CAN POISON!

Monika Davies

Gareth Stevens PUBLISHING

Please visit our website, www.garethstevens.com. For a free color catalog of all our high-quality books, call toll free 1-800-542-2595 or fax 1-877-542-2596.

Cataloging-in-Publication Data
Names: Davies, Monika.
Title: Plants can poison! / Monika Davies.
Description: New York : Gareth Stevens Publishing, 2023. | Series: Plants with superpowers | Includes glossary and index.
Identifiers: ISBN 9781538278994 (pbk.) | ISBN 9781538279014 (library bound) | ISBN 9781538279007 (6pack) | ISBN 9781538279021 (ebook)
Subjects: LCSH: Poisonous plants–Juvenile literature. | Plant defenses–Juvenile literature. Classification: LCC QK100.A1 D38 2023 | DDC 581.6'59–dc23

Published in 2023 by
Gareth Stevens Publishing
29 East 21st Street
New York, NY 10010

Editor: Monika Davies
Designer: Leslie Taylor

Portions of this work were originally authored by Dwayne Hicks and published as *Plants that Poison*. All new material in this edition authored by Monika Davies.

Photo credits: Cover, p.1 Leena Robinson/Shutterstock.com; series art (various accents) Ozz Design, Piotr Urakau, Designer things, Maxger/Shutterstock.com; p. 5 CampSmoke/Shutterstock.com; p. 7 (close-up) Brzostowska/Shutterstock.com; p. 7 Olha Solodenko/Shutterstock.com; p. 9 (buckthorn) Nelly B/Shutterstock.com; p. 9 (fly honeysuckle) ULD media/Shutterstock.com; p. 11 (both maps) https://www.cdc.gov/niosh/topics/plants/geographic.html; p. 11 (leaves) Kyle Selcer/Shutterstock.com; p. 11 (rash) Michael Moloney/Shutterstock.com; p. 13 (plant) Libin Jose/Shutterstock.com; p. 13 (jewelry) ireneuke/Shutterstock.com; p. 15 (plant) Ksenia Lada/Shutterstock.com; p. 15 (castor oil) Luis Echeverri Urrea/Shutterstock.com; p. 17 (hemlock) essicahyde/Shutterstock.com; p. 17 (stem) IanRedding/Shutterstock.com; p. 17 (queen anne's lace) Paul Reeves Photography/Shutterstock.com; p. 17 (flower) Jessica Hyde/Shutterstock.com; p. 19 (berries) Simon Groewe/Shutterstock.com; p. 19 (flower bell) Fabrizio Guarisco/Shutterstock.com; p. 21 Kendo Nice/Shutterstock.com.

Printed in the United States of America

CPSIA compliance information: Batch #CSGS23: For further information, contact Gareth Stevens, New York, New York, at 1-800-542-2595.

CONTENTS

Words in the glossary appear in **bold** type the first time they are used in the text.

Toxic Threats

Plants are a big part of what we eat every day. Fruits and vegetables often end up on our dinner plates. But did you know some plants can pose a toxic **threat**?

While many plants are safe to eat, some plants can poison you. These plants contain, or hold, toxins that can be deadly to humans. Just brushing up against a poisonous plant can sometimes cause you to get sick! Let's learn more about these plants with a toxic side.

When you go into nature, it helps to take a plant guidebook with you. It can tell you which plants are safe—and which to stay away from.

Staying Alive

Every plant faces dangers in their environment, or natural surroundings. Animals, bugs, and other plants can harm or kill them. Plants can't move to avoid, or stay way from, these dangers. Instead, they've had to adapt.

An adaptation is a change in a living thing that makes it better able to live in its surroundings. Making poison is an adaptation. Long ago, poisonous plants survived longer than nonpoisonous plants. They lived to pass down this adaptation. Over time, it became part of the **species**.

A poisonous plant is one with poison, which is something that can cause illness or death in another living being. An example of a poisonous plant is the water hemlock, which is pictured here.

On the Defense

Making poison is a plant defense, or way of guarding against an enemy. A plant's poison can greatly harm animals that eat or even touch it. Over time, predators have learned to stay away from poisonous plants.

Some plant defenses are easy to see. These plants have thorns, can look sick, or shrink when touched. However, poison is a chemical defense. Poison is a chemical made inside the plant, so it's hard to tell if a plant is poisonous just by looking at it.

Super Fact!

A chemical is matter that can be mixed with other matter to cause changes. A chemical defense is when a plant uses chemicals to keep from being eaten.

poisonous
buckthorn berries

poisonous
fly honeysuckle

Sometimes predators get unlucky and eat a poisonous plant. They often get very sick or even die.

A Poisonous Touch

Poison ivy is a well-known poisonous plant. It grows in many places, from the woods to your backyard. Poison ivy has leaves with three **leaflets**. The leaflets can be hairless or have saw-toothed edges.

When poison ivy is touched, it lets out a toxic oil called urushiol (oo-ROO-shee-awl). The oil causes an itchy, painful **rash** to appear on your skin. Sometimes, the reaction is so bad people have to go to the hospital. After one **encounter** with poison ivy, people avoid touching it, and the plant can grow without being bothered.

Super Fact!

The poison ivy plant releases urushiol from its leaves, flowers, fruit, stem, and roots. There's no safe way to handle this plant—the oil can even stick to your clothes for up to a year!

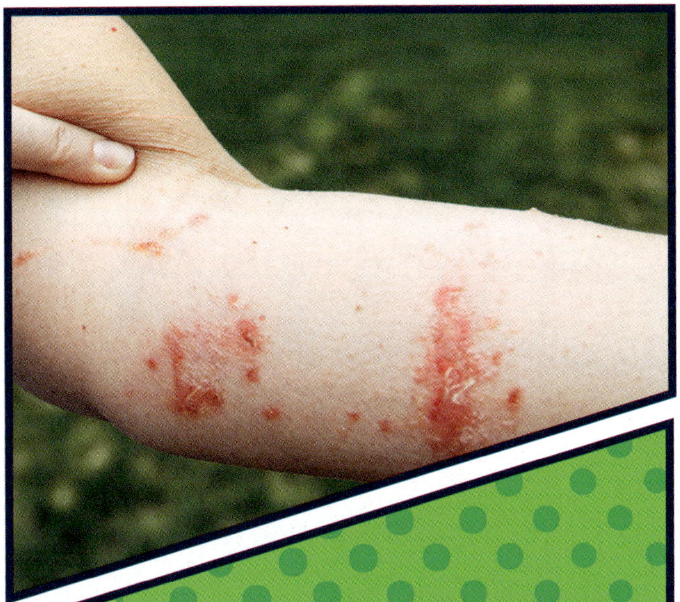

Besides Alaska and Hawaii, every U.S. state has one or more species of poison ivy, poison oak, or poison sumac. These all produce urushiol and cause rashes. Western poison ivy and eastern poison ivy are two common species.

eastern poison ivy locations

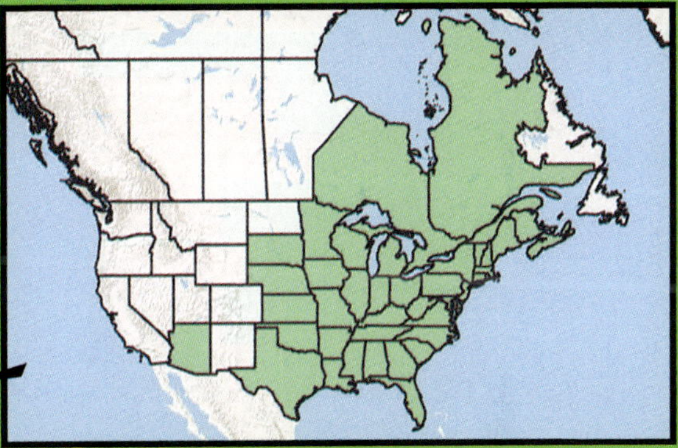

western poison ivy locations

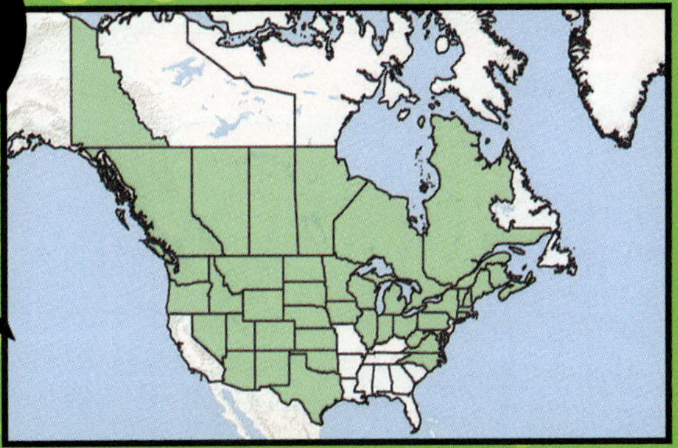

Fatal Beads

The jequirity plant is another plant with a toxic edge. It grows in warm, **tropical** places across the planet. The black-and-red seeds of this plant are used as beads in necklaces and to make instruments. This may seem surprising because eating even one of these poisonous seeds can be deadly!

Jequirity seeds contain the abrin toxin. People who have **ingested** this toxin will throw up, and inside body parts like the kidneys and liver may fail. They might then die after several days.

Super Fact!

If ingested, the abrin toxin finds its way into your body's cells and stops the cells from making **proteins**. This causes the cells to die.

Jequirity seeds have a hard coating on their outside that keeps the abrin toxin safely inside. If people swallow the seeds whole, they might not get sick.

Ricin Risks

Another **fatal** plant toxin is ricin. The castor oil plant contains deadly amounts of ricin in its seeds and smaller amounts in other parts of the plant. Poisoning happens when an animal eats or chews on broken seeds. The poison is let out, and the animal can get very sick and might even die.

Although castor seeds are highly poisonous, people have been using **medicine** made from castor oil for thousands of years. Once the oil is removed from the seeds, it's safe to take as medicine in small doses.

Super Fact!

It's believed that the castor oil plant first grew in warm areas in eastern Africa. Nowadays, the plant is mostly grown in India, China, and Brazil to make castor oil.

The castor oil plant can grow quite tall in warm conditions. In some places, the plant grows up to 40 feet (12.2 m) tall!

Wily Weeds

Poison hemlock is a wily, or tricky, weed that grows in wet soils found in sunny places. It is found along the edges of roads, fields, streams, and hiking trails. This poisonous weed has groups of tiny white flowers.

All parts of the poison hemlock plant are poisonous. It can kill animals and people who eat it. The poison contains a chemical that can shut down an animal's nervous system. This system carries messages from the brain to the rest of the body. That threat is enough to keep predators away!

Super Fact!

Poison hemlock can be mistaken for Queen Anne's lace. They look alike. Queen Anne's lace and poison hemlock both have hollow stems, but poison hemlock has a stem with tiny, purple spots.

Poison hemlock is part of the Apiaceae plant family. This plant family also includes carrots, celery, and parsley.

Poison hemlock has purple spots on its stems (below).

Queen Anne's lace looks like poison hemlock, but it's not poisonous.

Deadly Beauty

The belladonna plant's dark, bell-shaped flowers and shiny black berries make for an appealing picture. But for the unlucky people who've tasted the plant, the belladonna can spell death.

Belladonna is also called deadly nightshade. The entire plant is poisonous, but the sweet berries might present the most danger. People eat the berries, thinking they are a safe snack. People who ingest the poison from a belladonna plant can lose the ability to move their **muscles** and may even die.

The belladonna plant is part of a group of plants called nightshades. Eggplants, potatoes, and tomatoes are nightshades too. These plants are safe to eat though!

Look Twice!

The oleander plant has funnel-shaped flowers that catch the eye. Like other poisonous plants, every part of an oleander is toxic. Yet many people still grow the beautiful oleander in their home gardens. While the plant's poison can cause illness or even death, it is only toxic if eaten.

From far away, plants like the oleander look harmless. But make sure to look twice! Poisonous plants have toxins inside that you can't see, so it's best to stay away from unfamiliar plants while in nature.

Super Fact!

Deaths from oleander plants do not happen often. The plant is bitter to taste, and most people steer clear of eating it.

People like growing the oleander plant due to its striking blooms. The plant's flowers can be pink, red, white, or yellow.

GLOSSARY

encounter: A meeting or experience with something or someone.

fatal: Leading to death.

ingest: To take food or drink into the body.

leaflet: One of the parts of a larger leaf attached to a single stem.

medicine: A drug taken to make a sick person well.

muscle: One of the parts of the body that allow movement.

protein: A necessary element found in all living things.

rash: A group of red spots on the skin.

species: A group of plants or animals that are all of the same kind.

threat: Something likely to cause harm.

tropical: Having to do with the warm parts of Earth near the equator.

FOR MORE INFORMATION

BOOKS

Levy, Janey. *Poisonous Plants*. New York, NY: Gareth Stevens Publishing, 2020.

Schuh, Mari. *Poisonous Plants*. Minneapolis, MN: Jump! Inc., 2019.

WEBSITES

Plant Defense
www.dkfindout.com/us/animals-and-nature/plants/plant-defense/
Discover different kinds of plant defenses here.

The Secret World of Plants
www.cbc.ca/kidscbc2/the-feed/the-secret-world-of-plants
Learn more about plant tricks that help them survive.

INDEX